OnBoard
ACADEMICS

Time and Money

© 2015 OnBoard Academics, Inc
Portsmouth, NH
800-596-3175
www.onboardacademics.com
ISBN: 978-1-63096-074-2

OnBoard Academic's books are specifically designed to be used as printed workbooks or as on-screen instruction. Each page offers focused exercises and students quickly master topics with enough proficiency to move on to the next level.

OnBoard Academic's lessons are used in over 25,000 classrooms to rave reviews. Our lessons are aligned to the most recent governmental standards and are updated from time to time as standards change. Correlation documents are located on our website. Our lessons are created, edited and evaluated by educators to ensure top quality and real life success.

Interactive lessons for digital whiteboards, mobile devices, and PCs are available at www.onboardacademics.com. These interactive lessons make great additions to our books.

You can always reach us at customerservice@onboardacademics.com.

Time

Key Vocabulary

o'clock

minute hand

hour hand

colon

one-half

half past

The big hand and the little hand

What color is the minute hand? _____

What color is the hour hand? _____

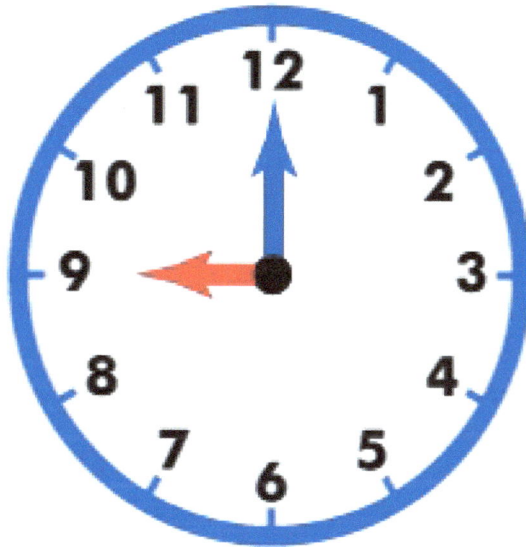

the time is nine o'clock
9:00

Time Facts

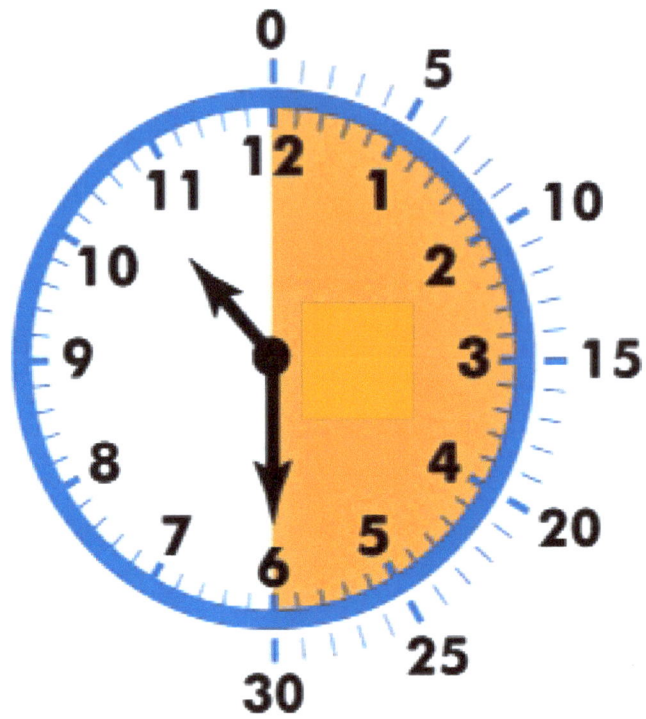

How many minutes in:

an hour   half an hour

Reading and Writing the time

Write the time in the boxes provided.

The time is four o'clock

The time is half past eight

What time is it?

Complete the chart below.

half past	1

8 : 00

6	o'clock

Set the clocks to match the times for Owen's favorite shows.
Draw in the minute and the hour hands.

10:00 AM	Cartoon Corner	
11:30 AM	Max and Muffin	
7:30 PM	Hero on the Hudson	

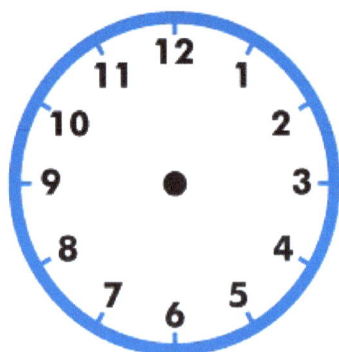

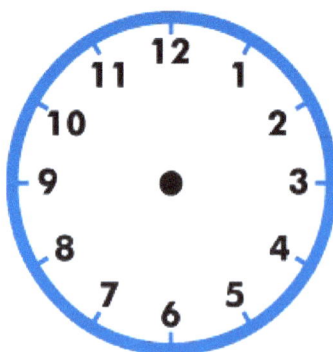

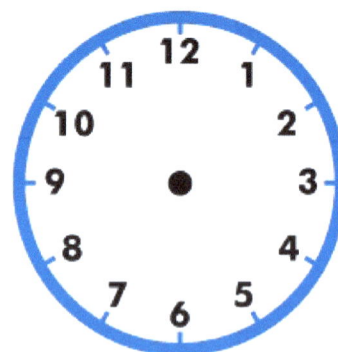

A.M. & P.M.

Fill in the time and make a check in the am or pm

AM starts at 12 o'clock at night
PM starts at 12 o'clock during the day

"Hurry, or you'll be late for school!"

		AM	
	. .	PM	

"We have one more lesson before lunch."

		AM	
	. .	PM	

"It's time for dismissal."

		AM	
	. .	PM	

Name_____

Time Quiz

Circle or fill in the correct answers.

1 **True or false? There are 12 minutes in half an hour.**

2 **What is the time?**

A 5:30

B 6:00

C 7:30

D 6:30

3 **How many minutes is it until 7:00?**

4 **What number is the minute hand pointing at when it is 9:00 PM?**

Time

Key Vocabulary

A.M.

P.M.

Midnight

Noon

Elapsed time

What time is it?

Put a check mark in the box next to the correct time.

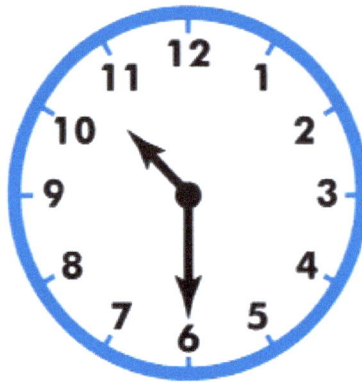

☐	quarter to two	☐	5:53	☐	11:40
☐	9:08	☐	10:30	☐	7:58
☐	2.45	☐	eleven thirty	☐	11:20

A.M., P.M., Midnight or Noon

Write the proper time and midnight or noon under the correct illustration.

Write the time and then decide if its A.M. or P.M. based on the clue.

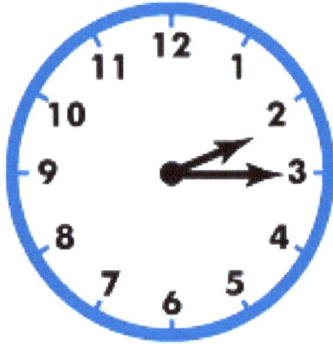

"It's the end of the school day."

"I'm fast asleep in bed."

"It's time for lunch!"

How long do we get for lunch.

Start time

End time

Elapsed time

Time and Elapsed Time

Write the times and then the elapsed times for each activity.

Home from school until bedtime

Times:

Total Elapsed Time:

Homework

End Time:

Elapsed Time:

 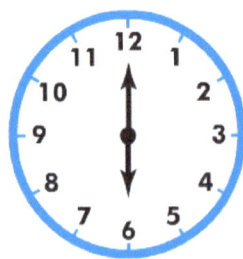

Computer/TV

End Time:

Elapsed Time:

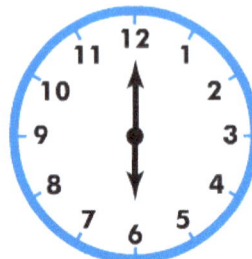 

Dinner

End Time:

Elapsed Time:

Name_____

Time Quiz

1 True or false? The time at midnight is 12:00 AM.

2 If it is 11:40 AM, what time will it be in 35 minutes?

  A 12:15 PM

  B 12:15 AM

  C 1:15 AM

  D 1:15 PM

3 What is the elapsed time (in minutes) from 1:50 PM until 2:15 PM?

4 What is the elapsed time (in minutes) from 11:55 PM until 12:00 AM?

Days, Weeks, Months, Years

Key Vocabulary

Days of the week

Months of the year

Can you order and count the days of the week?

Write the day in the proper box and add its day number for the week.

1	Sunday
3	Tuesday
5	Thursday

Friday

Monday

Saturday

Wednesday

Days of the Week Word Search

B	A	S	U	N	D	A	Y	S	D	Y
C	F	T	U	B	I	G	A	V	X	Z
C	W	E	D	N	E	S	D	A	Y	Y
Y	A	D	F	M	P	R	S	A	Y	Y
A	T	X	A	Q	U	A	E	A	Y	Y
D	N	N	P	G	T	K	U	L	I	F
N	W	P	M	U	N	O	T	Z	E	R
O	A	A	R	D	A	Y	Y	A	C	I
M	W	D	V	U	T	A	B	H	K	D
M	A	D	M	U	T	A	T	A	Y	A
Y	R	S	T	H	U	R	S	D	A	Y

Sunday

Monday

Tuesday

Wednesday

Thursday

Friday

Saturday

Circle the months that are not in the proper order.

JANUARY
 	
S M T W T F S
    1 2 3 4 5
6 7 8 9 10 11 12
13 14 15 16 17 18 19
20 21 22 23 24 25 26
27 28 29 30 31
	FEBRUARY
	
	S M T W T F S
	        1 2
	3 4 5 6 7 8 9
	10 11 12 13 14 15 16
	17 18 19 20 21 22 23
	24 25 26 27 28 29
		MARCH
		
		S M T W T F S
		          1
		2 3 4 5 6 7 8
		9 10 11 12 13 14 15
		16 17 18 19 20 21 22
		23 24 25 26 27 28 29
		30 31

APRIL S M T W T F S 1 2 3 4 5 6 7 8 9 10 11 12 13 14 15 16 17 18 19 20 21 22 23 24 25 26 27 28 29 30		

MAY
 	
S M T W T F S
        1 2 3
4 5 6 7 8 9 10
11 12 13 14 15 16 17
18 19 20 21 22 23 24
25 26 27 28 29 30 31
	JUNE
	
	S M T W T F S
	1 2 3 4 5 6 7
	8 9 10 11 12 13 14
	15 16 17 18 19 20 21
	22 23 24 25 26 27 28
	29 30
		AUGUST
		
		S M T W T F S
		          1 2
		3 4 5 6 7 8 9
		10 11 12 13 14 15 16
		17 18 19 20 21 22 23
		24 25 26 27 28 29 30
		31

JULY S M T W T F S 1 2 3 4 5 6 7 8 9 10 11 12 13 14 15 16 17 18 19 20 21 22 23 24 25 26 27 28 29 30 31		

SEPTEMBER
 	
S M T W T F S
1 2 3 4 5 6
7 8 9 10 11 12 13
14 15 16 17 18 19 20
21 22 23 24 25 26 27
28 29 30
	NOVEMBER
	
	S M T W T F S
	            1
	2 3 4 5 6 7 8
	9 10 11 12 13 14 15
	16 17 18 19 20 21 22
	23 24 25 26 27 28 29
	30
		OCTOBER
		
		S M T W T F S
		      1 2 3 4
		5 6 7 8 9 10 11
		12 13 14 15 16 17 18
		19 20 21 22 23 24 25
		26 27 28 29 30 31

DECEMBER S M T W T F S 1 2 3 4 5 6 7 8 9 10 11 12 13 14 15 16 17 18 19 20 21 22 23 24 25 26 27 28 29 30 31		

Seasons

In Boston, Massachusetts the year has four very different seasons. In summer it is sunny and hot, in winter it snows and is very cold, in spring the flowers bloom and in the fall the trees lose their leaves before winter.

On the calendar below draw a line from the highlighted month with the icon that best represents the weather for that month.

JANUARY

S	M	T	W	T	F	S
		1	2	3	4	5
6	7	8	9	10	11	12
13	14	15	16	17	18	19
20	21	22	23	24	25	26
27	28	29	30	31		

FEBRUARY

S	M	T	W	T	F	S
					1	2
3	4	5	6	7	8	9
10	11	12	13	14	15	16
17	18	19	20	21	22	23
24	25	26	27	28	29	

MARCH

S	M	T	W	T	F	S
						1
2	3	4	5	6	7	8
9	10	11	12	13	14	15
16	17	18	19	20	21	22
23	24	25	26	27	28	29
30	31					

APRIL

S	M	T	W	T	F	S
		1	2	3	4	5
6	7	8	9	10	11	12
13	14	15	16	17	18	19
20	21	22	23	24	25	26
27	28	29	30			

MAY

S	M	T	W	T	F	S
				1	2	3
4	5	6	7	8	9	10
11	12	13	14	15	16	17
18	19	20	21	22	23	24
25	26	27	28	29	30	31

JUNE

S	M	T	W	T	F	S
1	2	3	4	5	6	7
8	9	10	11	12	13	14
15	16	17	18	19	20	21
22	23	24	25	26	27	28
29	30					

JULY

S	M	T	W	T	F	S
		1	2	3	4	5
6	7	8	9	10	11	12
13	14	15	16	17	18	19
20	21	22	23	24	25	26
27	28	29	30	31		

AUGUST

S	M	T	W	T	F	S
					1	2
3	4	5	6	7	8	9
10	11	12	13	14	15	16
17	18	19	20	21	22	23
24	25	26	27	28	29	30
31						

SEPTEMBER

S	M	T	W	T	F	S
	1	2	3	4	5	6
7	8	9	10	11	12	13
14	15	16	17	18	19	20
21	22	23	24	25	26	27
28	29	30				

OCTOBER

S	M	T	W	T	F	S
			1	2	3	4
5	6	7	8	9	10	11
12	13	14	15	16	17	18
19	20	21	22	23	24	25
26	27	28	29	30	31	

NOVEMBER

S	M	T	W	T	F	S
						1
2	3	4	5	6	7	8
9	10	11	12	13	14	15
16	17	18	19	20	21	22
23	24	25	26	27	28	29
30						

DECEMBER

S	M	T	W	T	F	S
	1	2	3	4	5	6
7	8	9	10	11	12	13
14	15	16	17	18	19	20
21	22	23	24	25	26	27
28	29	30	31			

Reading a Calendar
Use what you know about days of the week to complete this exercise.

JANUARY

S	M	T	W	T	F	S
		1	2	3	4	5
6	7	8	9	10	11	12
13	14	15	16	17	18	19
20	21	22	23	24	25	26
27	28	29	30	31		

(S) What day of the week is this?   **(T)** What about this?

What day of the week is January 11?

What is the date of the third Wednesday?

How many Mondays are there in this January?

This is the calendar for 2036. The last time the calendar was exactly like this was 2008.

2036

JANUARY

S	M	T	W	T	F	S
		1	2	3	4	5
6	7	8	9	10	11	12
13	14	15	16	17	18	19
20	21	22	23	24	25	26
27	28	29	30	31		

FEBRUARY

S	M	T	W	T	F	S
					1	2
3	4	5	6	7	8	9
10	11	12	13	14	15	16
17	18	19	20	21	22	23
24	25	26	27	28	29	

MARCH

S	M	T	W	T	F	S
						1
2	3	4	5	6	7	8
9	10	11	12	13	14	15
16	17	18	19	20	21	22
23	24	25	26	27	28	29
30	31					

APRIL

S	M	T	W	T	F	S
		1	2	3	4	5
6	7	8	9	10	11	12
13	14	15	16	17	18	19
20	21	22	23	24	25	26
27	28	29	30			

MAY

S	M	T	W	T	F	S
				1	2	3
4	5	6	7	8	9	10
11	12	13	14	15	16	17
18	19	20	21	22	23	24
25	26	27	28	29	30	31

JUNE

S	M	T	W	T	F	S
1	2	3	4	5	6	7
8	9	10	11	12	13	14
15	16	17	18	19	20	21
22	23	24	25	26	27	28
29	30					

JULY

S	M	T	W	T	F	S
		1	2	3	4	5
6	7	8	9	10	11	12
13	14	15	16	17	18	19
20	21	22	23	24	25	26
27	28	29	30	31		

AUGUST

S	M	T	W	T	F	S
					1	2
3	4	5	6	7	8	9
10	11	12	13	14	15	16
17	18	19	20	21	22	23
24	25	26	27	28	29	30
31						

SEPTEMBER

S	M	T	W	T	F	S
	1	2	3	4	5	6
7	8	9	10	11	12	13
14	15	16	17	18	19	20
21	22	23	24	25	26	27
28	29	30				

OCTOBER

S	M	T	W	T	F	S
			1	2	3	4
5	6	7	8	9	10	11
12	13	14	15	16	17	18
19	20	21	22	23	24	25
26	27	28	29	30	31	

NOVEMBER

S	M	T	W	T	F	S
						1
2	3	4	5	6	7	8
9	10	11	12	13	14	15
16	17	18	19	20	21	22
23	24	25	26	27	28	29
30						

DECEMBER

S	M	T	W	T	F	S
	1	2	3	4	5	6
7	8	9	10	11	12	13
14	15	16	17	18	19	20
21	22	23	24	25	26	27
28	29	30	31			

What day will your birthday be on in the year 2036?

How old will you be then?

How many?
Circle the answer.

Days in a week	8	7	31	25
Months in a year	11	12	13	15
Weeks in a year	31	12	52	36
Days in a year	365	52	500	100
Days in January	31	32	39	29

Name_____

Days, Weeks, Months, Years Quiz

Circle of fill in the correct answer.

1 **Which month comes after July?**

A September B August C June D March

2 **In which month is Thanksgiving?**

A November B October C June D April

3 **How many days are there in December?**

4 **How many days are there in 3 weeks?**

Money

Key Vocabulary

penny

nickel

dime

quarter

half-dollar

Which one am I?

Draw a line from the coin to the circle under the coin's name. Write the "cents" into the box.

nickel	quarter	dime	half-dollar	penny
◯	◯	◯	◯	◯

| 1 cent | 5 cents | 25 cents | 50 cents | 10 cents |

How much did you save?
Write the amount under each piggy bank.

Put the correct amount in each box.

You can try to draw the coins or just write the name of the coin and how many. For example you can write Five Quarters.

52 cents	80 cents

77 cents	16 cents

I spent this much money, what did I buy.
Add the coins and then circle the item that you bought.

31 cents

51 cents

56 cents

65 cents

80 cents

60 cents

Who has the most money - Fernando or Mia?
Add the allowance and the savings and write your answer in the space provided. After you have added the allowance and savings add them together for a total. What coins do you have once the allowance and savings are added together. Put a check mark next to the person with the highest amount.

	Allowance	Savings	Total
Fernando	cents	cents	cents
Mia	cents	cents	cents

Grandmother has agreed to give each child enough money so that they have $1.00.
How much does Grandmother need to add to each child's savings?

	Savings	Grandmother
Tori		
	cents	cents
Owen		
	cents	cents

Name_____

Money Quiz
Circle of fill in the correct answer.

1 **True or false? A dime is worth 5 cents.**

2 **If I have 2 quarters, 1 nickel and 1 penny, I have:**

- **A** **25 cents**

- **B** **61 cents**

- **C** **56 cents**

- **D** **31 cents**

3 **How many cents do I have?**

4 **= how many nickels?**

Money

Key Vocabulary

Coin

Bill

Decimal(s)

How many of these are in those?

 in

 in

 in

 in

 in

Hint:
Penny is 1¢
Nickel is 5¢
Dime is 10¢
Quarter is 25¢
Dollar is 100¢

Write the amount in cents and in dollars.

¢   ¢   ¢

$ $ $

Write these amounts in dollars.

$

$

$

Use the 'counting on' method to make change.
Draw the coins in the box in order for 'counting on.'

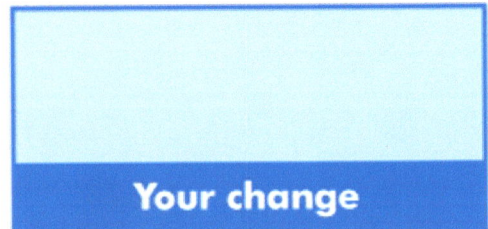

84¢	You pay with this	Your change

Counting On Method

$$84 + 1 = 85$$
$$85 + 5 = 90$$
$$90 + 10 = 100$$
$$\underline{16}$$

Make change for these purchases.

79¢	You pay with this	$ _____ Your change
$1.26	You pay with this	$ _____ Your change
$3.85	You pay with this	$ _____ Your change
$14.49	You pay with this	$ _____ Your change

Name_____

Money Quiz

Fill in or circle the correct answer.

(1) **True or false? Two dimes is the same amount as four nickels.**

(2) **What is the total value of these coins?**

    (A) **82¢**

    (B) **97¢**

    (C) **$0.92**

    (D) **$1.02**

(3) **If you purchase a soda for $1.27, how much change would you receive from two dollars ? (Answer as a decimal).**

(4) **If you purchase a 'Super Meal' for $3.69, how much change would you receive from $10? (Answer as a decimal).**

www.ingramcontent.com/pod-product-compliance
Lightning Source LLC
Chambersburg PA
CBHW052048190326
41521CB00002BA/152